AI AWARENESS SERIES

AI in Construction and Engineering

Darian Batra

Contents

Introduction

Artificial intelligence is rapidly reshaping the construction and engineering industries—revolutionizing how we design, plan, build, and manage the spaces and infrastructure we rely on every day. What was once seen as futuristic is now becoming foundational. From smart buildings and automated equipment to intelligent project planning and predictive maintenance, AI is transforming every phase of the construction lifecycle.

This book, AI in Construction and Engineering, is part of the AI Awareness Series—a collection designed to make artificial intelligence accessible, actionable, and responsible across real-world industries. Here, our focus is on one of the most complex and high-stakes sectors: the built environment.

You don't need to be a data scientist or software developer to benefit from what AI offers. This book is written for professionals across the field: architects and engineers, site managers and planners, construction executives, and policymakers. Whether you're just beginning to explore AI or already implementing smart technologies in your projects, this guide aims to demystify the tools and use cases shaping the industry's future.

The goal of this book is to provide both foundational understanding and practical insight. You'll see not only what's possible, but how AI can be applied thoughtfully and effectively within your own context.

Welcome to AI in Construction and Engineering. Let's explore how intelligent technologies are shaping smarter design, safer builds, and more resilient infrastructure.

Chapter 1: Foundations of AI in Design and Construction

Let's start by understanding generative design. This approach uses algorithms to explore a vast number of design possibilities quickly and efficiently. It's not just about speed—these algorithms also help optimize the performance of buildings, whether structurally, environmentally, or both. By enabling rapid iteration and supporting creative exploration, generative design opens the door to more innovative solutions in architecture and engineering.

Now, let's review the fundamentals of Building Information Modeling, or BIM. BIM creates a detailed digital representation of a building, capturing both its physical attributes and functional characteristics. This digital model allows for accurate visualization and facilitates collaboration among architects, engineers, and contractors within a shared environment. On top of that, BIM supports project management by integrating data and processes, helping teams coordinate more effectively throughout the project lifecycle.

When we bring generative design and BIM together, we unlock some powerful advantages—but also face certain challenges. On the plus side, this integration allows teams to explore design solutions more efficiently and improve overall project collaboration by centralizing data and workflows. However, there can be hurdles, such as software compatibility issues and the need for workflow adjustments. Adopting these technologies often requires changes in processes and sometimes additional training.

Artificial intelligence is also making a big impact on 3D modeling. AI can automate complex calculations, dramatically reducing manual effort and speeding up the design process. Beyond that, AI is able to recognize recurring design patterns, which helps streamline model creation and ensure consistency. By analyzing data, AI can even suggest improvements, making the overall design more efficient and of higher quality.

One of the most practical applications of AI is in automating repetitive design tasks. Tools powered by AI can take over routine activities like drafting and checking for errors, freeing up designers to focus on more creative and complex aspects of their work. This shift not only boosts productivity but also encourages innovative thinking by reducing the time spent on mundane tasks.

AI-driven 3D modeling isn't just theory—it's already being used in real-world projects across architecture and engineering. These case studies demonstrate how AI is helping teams deliver more efficient, accurate, and creative design solutions. By adopting AI tools, firms can stay competitive and meet the increasing demands of modern construction projects.

Let's look at how clash detection works traditionally in BIM projects. Usually, this involves manual reviews, where people painstakingly inspect models for conflicts. Some tools use rule-based software to automate parts of this process, but these methods can still miss complex issues. Overall, traditional clash detection can be slow and prone to human error, which increases the risk of costly mistakes during construction.

With machine learning, clash detection becomes much more dynamic. Algorithms continuously analyze the design data in real time, detecting potential conflicts as the model evolves. This proactive approach means clashes can be identified and addressed much earlier in the process, which helps avoid expensive errors and accelerates project timelines. It also promotes smoother collaboration among team members.

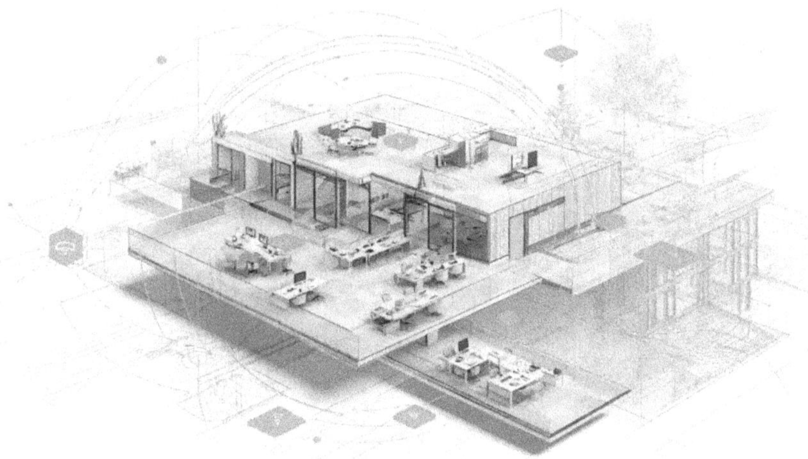

Real-time clash detection has a significant impact on project efficiency and error reduction. By identifying conflicts early, teams can prevent costly rework later in the project. This also shortens the overall project schedule, since delays caused by late-stage corrections are minimized. Plus, the collaborative environment improves as stakeholders work together using shared digital tools and receive timely updates on issues that need resolving.

To wrap up, the integration of generative design, AI-driven 3D modeling, and machine learning-powered clash detection is reshaping the way we approach design and construction. These technologies are not just about making processes faster—they're about enhancing creativity, collaboration, and efficiency. As we adopt these innovations, we're setting the stage for smarter, more effective workflows in the built environment.

Chapter 2: Data as the Foundation

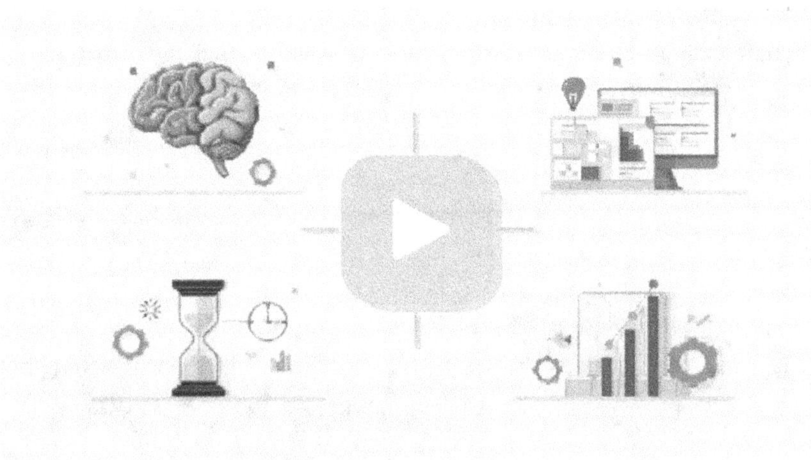

Predictive scheduling uses data-driven forecasting to anticipate project milestones and risks with greater accuracy. By analyzing past data and trends, we can create models that help us predict future outcomes. Simulation techniques allow us to test different scenarios and understand how various factors impact the schedule. Machine learning models take this a step further by analyzing large datasets to uncover patterns we might miss, improving both our predictions and the accuracy of our project timelines. And finally, historical trend analysis supports continuous improvement in how we plan and manage risks.

Compared to traditional scheduling methods, predictive project scheduling offers several key benefits. It improves accuracy by using advanced forecasting methods, making our timelines more reliable. It also supports proactive risk management, helping us identify and address potential problems before they escalate. Resource utilization is another major advantage—predictive scheduling ensures that resources are allocated efficiently, boosting productivity and reducing waste. Finally, it gives us increased flexibility by allowing schedules to adapt dynamically to changes, making projects easier to manage in complex environments.

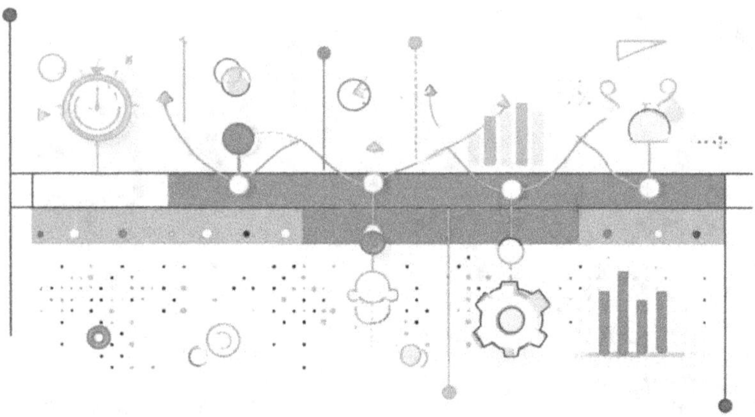

Let's look at some of the key tools and technologies that make predictive scheduling possible. AI-powered software helps automate tasks and enhance decision-making with intelligent algorithms. Predictive analytics platforms allow us to forecast future trends and optimize project plans. And advanced project management systems offer real-time monitoring and the flexibility to adjust schedules and resources as projects evolve. Together, these tools form the backbone of a modern, predictive project management approach.

Identifying potential delays is a critical part of project success. We can use data analysis—both historical and real-time—to spot trends and early warning signs of delays. Expert judgment is also vital, drawing on the experience of seasoned professionals to assess risk factors. Scenario analysis allows us to explore different possibilities and plan for them. And AI algorithms give us a powerful tool for detecting patterns and anomalies, helping us catch potential delays early and make better decisions.

When it comes to assessing risks, we can use both quantitative and qualitative methods. Quantitative risk assessment often involves tools like Monte Carlo simulations, which use statistical modeling to predict delay risks with a high degree of accuracy. Qualitative risk assessment focuses on insights gained from expert interviews and judgment. By combining both approaches, we can get a more accurate and well-rounded view of potential risks and how to manage them.

Integrating forecasting into project timelines helps us manage risks proactively. By embedding risk forecasts directly into the schedule, we can anticipate and respond to delays more effectively. This also allows for timely plan adjustments and strategic buffer allocation, giving us the flexibility to handle uncertainties. And by keeping stakeholders informed with clear communication about risks and schedule changes, we can build trust and support throughout the project.

AI algorithms can greatly enhance how we allocate project resources. Machine learning optimization analyzes data patterns to distribute resources efficiently under various constraints. Genetic algorithms mimic natural selection to evolve and refine resource allocation strategies. And reinforcement learning improves resource management by learning from feedback and adjusting over time, making it particularly useful in dynamic project environments.

AI also supports real-time decision-making. Continuous monitoring systems keep track of project metrics, providing up-to-date information that helps guide decisions. AI-powered adaptive recommendations suggest ways to optimize resources and timelines on the fly. And dynamic resource adjustment means project managers can make changes in real time based on AI insights, leading to better project outcomes.

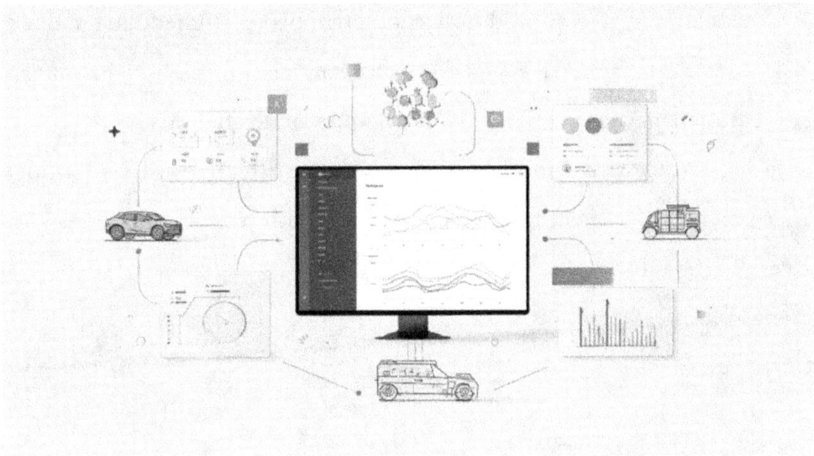

We've seen some compelling real-world results from AI-assisted resource allocation. AI has driven significant efficiency gains across various industries by optimizing how resources are used. This often translates into real cost savings, thanks to reduced waste and better utilization. AI-powered resource management has also improved schedule adherence, helping projects stay on track through smarter allocation and continuous monitoring.

To wrap up, we've explored how predictive project scheduling, delay risk forecasting, and AI-assisted resource allocation are transforming project management. These advanced approaches help us work smarter, reduce risks, and use resources more effectively. By adopting these methods, project teams can improve both project outcomes and overall efficiency.

Chapter 3: Building an AI-Ready Organisation

Computer vision technologies are transforming construction site monitoring. Using AI-powered analysis, these systems can automatically assess site conditions with accuracy and efficiency. Cameras placed around the site help detect on-site activities, which supports both progress tracking and safety management — all without relying on constant manual supervision. And importantly, continuous data collection means project teams get real-time insights into site status and construction progress, helping them make better decisions faster.

Automated site monitoring brings significant benefits. First, it improves accuracy by reducing human error and minimizing the need for manual data collection. Second, it enhances safety oversight. With constant monitoring and real-time alerts, the system helps reduce the risk of accidents on-site. Finally, by providing continuous data streams, these systems enable better decision-making and more efficient allocation of resources throughout a project.

One of the strengths of computer vision is its ability to integrate with existing site management systems. This integration allows project teams to optimize workflows and improve coordination across different tools. For safety, it means real-time monitoring with automated alerts and analysis, improving overall site safety. And with efficient data exchange between systems, teams can make better-informed decisions and manage sites more effectively.

A key safety feature of computer vision is real-time detection of personal protective equipment, or PPE. AI-powered cameras continuously monitor workers to ensure they are wearing essential gear like helmets and high-visibility vests. If the system spots someone not following PPE requirements, it can immediately alert supervisors — helping to enforce safety rules and protect workers.

Beyond PPE, computer vision helps identify hazardous behaviors and risky areas on-site. It can detect unauthorized access or unsafe movements, which helps prevent potential accidents before they happen. The system also recognizes dangerous zones automatically, sending alerts to keep personnel informed and reduce risk. By detecting these issues early, project teams can take proactive steps to improve safety and prevent incidents.

Drones and static cameras play a big role in collecting data for progress tracking. Drones provide aerial coverage, giving a comprehensive view of the construction site from multiple angles and heights. At the same time, static cameras keep watch over key locations, offering continuous, fixed-point monitoring. By combining data from both drones and static cameras, teams get a richer, more complete picture of the site for analysis and decision-making.

One powerful use of computer vision is in comparing as-built conditions with as-planned models. Advanced algorithms process images to identify any differences between what's been built and the original plans. Detecting deviations early allows teams to correct mistakes before they become costly or cause project delays. This constant monitoring helps ensure that construction stays on schedule and meets the required specifications.

Visualization and reporting tools are critical for keeping project stakeholders informed. Computer vision systems often come with features that help visualize data and generate clear reports. These tools make it easier to communicate progress, highlight potential issues, and support transparent project management — all of which contribute to better stakeholder engagement and decision-making.

To wrap up, computer vision is proving to be a game-changer in construction site monitoring. It enhances safety, improves progress tracking, and supports better decision-making through real-time data and advanced analysis tools. By integrating seamlessly with existing systems, it helps construction teams manage sites more efficiently and safely — making it a valuable asset for modern construction projects.

Chapter 4: AI Project Implementation

Robotics in construction is a rapidly growing field. Automated bricklaying robots can lay bricks with high precision, which not only speeds up the process but also ensures consistent quality. Robotic welding systems are enhancing the accuracy and safety of metal joining tasks, reducing the risk of human error and workplace accidents. And when it comes to moving materials around a site, automated transportation systems are helping reduce manual labor, improve efficiency, and minimize delays.

There are several key benefits of introducing automation on construction sites. First, automation improves safety by taking over hazardous tasks and reducing risks to human workers. Second, machines can operate continuously without fatigue, which means projects get completed faster, boosting overall productivity. Third, automation enhances the accuracy and consistency of construction work, minimizing mistakes and leading to higher-quality outcomes.

However, deploying autonomous machinery isn't without challenges. The initial cost of these machines is high, which can make it a tough budget decision for many companies. They also require skilled operators who can manage and oversee their use safely and effectively. Integrating these machines into existing workflows isn't always straightforward and can create logistical hurdles. And finally, many autonomous systems struggle to adapt to complex or unpredictable site conditions, which limits their flexibility.

AI-powered earthmoving equipment relies on several technologies. Sensors play a critical role by gathering real-time data on terrain and obstacles, helping machines operate accurately. Machine learning algorithms allow these machines to analyze conditions and make autonomous decisions to optimize excavation tasks. GPS navigation ensures that the equipment can move and perform its tasks precisely without constant human input.

In practice, AI-guided earthmoving equipment brings several real-world benefits. Safety is enhanced because AI reduces operator fatigue and limits their exposure to dangerous conditions. Efficiency improves because AI optimizes routes and fuel usage, making operations more streamlined. Productivity also gets a boost, as these machines can work continuously without breaks, significantly speeding up construction projects.

In prefabrication, collaborative robots—also known as cobots—are making a big impact. They assist with assembly tasks by performing repetitive and precision work efficiently and consistently. Unlike traditional industrial robots, cobots are built with safety features that allow them to work alongside human workers, making them ideal for shared workspaces.

Cobots are designed to enhance human labor rather than replace it. By handling physically demanding or repetitive tasks, cobots reduce the strain on human workers and improve productivity. When integrated into workflows, cobots help streamline processes, improve coordination, and reduce the likelihood of errors during assembly, leading to more efficient operations overall.

The use of cobots in prefab construction also impacts quality, cost, and project timelines. Their precision enhances product quality by minimizing errors. They help reduce costs by cutting down on waste and labor expenses. And because they speed up the assembly process, they contribute to shorter project timelines, allowing manufacturers and builders to meet deadlines more easily.

To wrap up, robotics and autonomous machinery are driving significant advancements in the construction industry. While there are challenges to adopting these technologies, the potential benefits in terms of safety, productivity, cost savings, and quality make them a powerful tool for modern construction projects.

Chapter 5: Smart Estimating And Cost Control

Let's start with traditional estimating practices. These typically involve manual quantity takeoffs, where estimators calculate materials by hand from drawings. They also rely on cost databases with historical pricing data. While these methods have been standard in the industry, they are time-consuming and prone to human error, making them less adaptable to changes that occur during a project.

Now, let's talk about the integration of digital tools in cost estimation. Building Information Modeling, or BIM, offers detailed 3D models that improve both accuracy and visualization. Estimation software automates many calculations, reducing errors and speeding up the entire process. Cloud collaboration platforms also make a big difference by allowing real-time updates and easier teamwork during cost estimation projects.

Smart estimating brings several key benefits. First, it enhances precision, reducing errors and uncertainties in project cost predictions. It also helps lower the risk of cost overruns by improving forecasting and risk management. With smart estimating, the estimating cycles are much faster, allowing quicker decision-making. Finally, it supports dynamic updates, scenario analysis, and increases transparency in both budgeting and forecasting.

Let's move on to forecasting material price trends. Several key factors influence material prices. Supply chain disruptions can delay shipments and create scarcity, which drives prices up. Geopolitical events—like conflicts or trade restrictions—also impact costs. Market demand and currency fluctuations play a big role, as do changes in the availability of raw materials. All of these factors combine to make price forecasting a critical skill.

To forecast prices, we rely on a mix of techniques and tools. Statistical analysis, time series modeling, and machine learning are common forecasting techniques. Tools like price indices, commodity market data, and specialized software give us the ability to predict future price trends with greater accuracy.

Predictive analytics is especially useful for procurement planning. It helps us anticipate future price changes, enabling better purchasing decisions. By using data-driven insights, we can time purchases to get the best prices, optimizing cost efficiency. It also supports scenario planning, helping reduce financial risks by preparing for different market outcomes.

Now, let's examine change orders and their challenges. A change order is essentially a modification to the contract, and these often cause project delays and increased costs. Assessing their impact is complex—it takes careful analysis to understand how they'll affect the project timeline and budget. Also, coordinating among stakeholders can be tricky, since everyone has different priorities and clear communication is crucial.

Artificial Intelligence plays a powerful role in impact assessment. AI can analyze large amounts of data to evaluate change orders and predict their impact. Using natural language processing, or NLP, AI can even understand and process the text within change orders. AI also forecasts the cost and schedule impacts of these changes, giving project managers the insights they need to make better decisions. Ultimately, AI delivers actionable insights that help teams manage changes effectively and reduce risk.

There are already real-world cases showing how AI improves change order analysis. AI tools increase the accuracy of analyzing change orders, cutting down on documentation errors. They also speed up the change order process through automation, which helps reduce project delays. Accurate AI analysis means fewer disputes between project stakeholders. Overall, AI enhances project control and decision-making when it comes to managing change orders and their costs.

To conclude, advanced techniques like smart estimating, material price forecasting, and AI-driven change order analysis are changing the way we manage construction costs. These innovations help improve project accuracy, reduce risks, and enhance decision-making. By adopting these methods, you can deliver projects more efficiently and maintain stronger control over costs and timelines.

Chapter 6: Procurement And Supply Chain Intelligence

Procurement and supply chain intelligence is about using data-driven insights to manage and optimize the supply chain. It starts with collecting accurate, timely data from every part of the supply chain — from suppliers to customers. This intelligence gives organizations better visibility and agility, allowing them to react quickly to market changes or disruptions. Ultimately, leveraging this intelligence helps companies make smarter procurement decisions and boosts overall efficiency.

Several key technologies enable supply chain intelligence today. Artificial intelligence and machine learning analyze complex data patterns, helping predict future trends and optimize decisions. The Internet of Things connects devices across the supply chain, providing real-time monitoring and data collection. Big data analytics processes huge amounts of information to support predictive analysis and smarter decision-making. And finally, decision-support tools bring all this data together to help managers make well-informed supply chain choices.

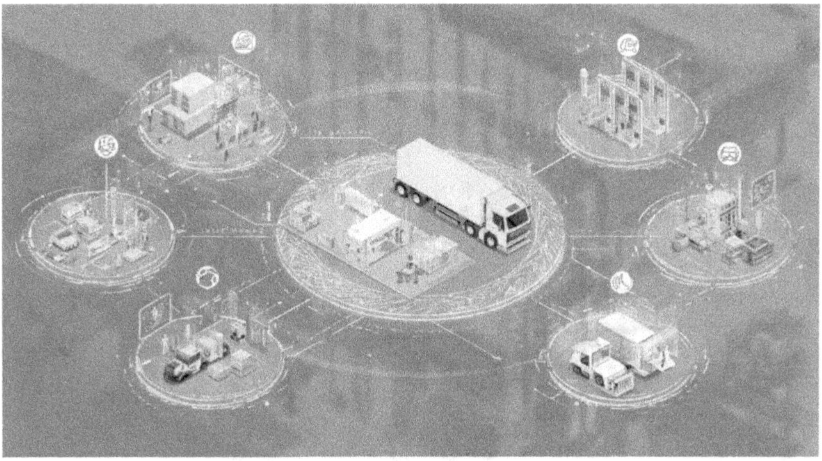

There are clear benefits when modern procurement strategies embrace supply chain intelligence. First, efficiency improves as processes become streamlined and delays are reduced. Second, enhanced risk management helps businesses identify and mitigate risks before they become major issues. Third, cost reduction — by analyzing spending and optimizing procurement — leads to significant savings. And finally, supplier collaboration improves, strengthening relationships and enhancing communication with key suppliers.

Supplier risk prediction starts with identifying key risk factors that could impact supplier relationships. Financial instability can disrupt operations and affect your supply chain. Geopolitical issues — like trade restrictions or political unrest — may affect a supplier's ability to deliver. Quality problems can lead to defective products and harm customer satisfaction. And delivery delays can bring production to a halt and result in lost revenue.

To accurately assess supplier risk, we rely on diverse data sources — financial reports, market trends, even social media insights and historical records. Advanced analytics techniques are then used to process this data, helping predict risks and support better decision-making in procurement.

Integrating risk prediction into procurement decisions offers several advantages. It supports informed supplier selection by highlighting potential vulnerabilities. It enables proactive contingency planning, ensuring businesses can respond effectively to disruptions. And it enhances negotiation strategies, allowing businesses to secure better terms and reduce procurement risks.

Just-in-Time, or JIT delivery, is built on the principle of aligning inventory supply closely with actual demand. The goal is to reduce excess inventory and minimize storage costs, ensuring products arrive precisely when needed — not too early, and not too late.

Accurate demand forecasting is crucial for JIT delivery. We use statistical analysis to identify patterns from historical data. Machine learning models take this further by learning from data and continuously improving their predictions. Real-time data — like current sales trends, market conditions, and customer behavior — also feeds into these models, helping forecasts stay relevant and accurate.

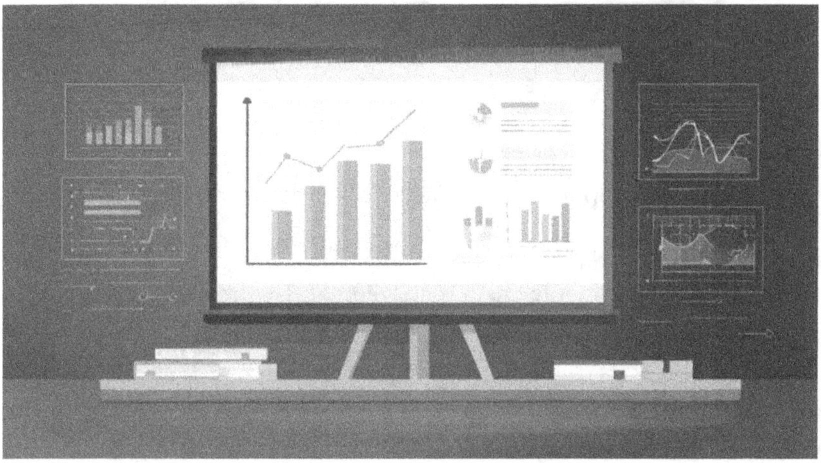

Implementing JIT forecasting comes with its challenges. Data quality, variability in supply, and coordination complexity can make it tough to get right. But the benefits are significant — reduced inventory holding costs, improved cash flow, and better customer satisfaction thanks to on-time delivery and fewer delays.

In conclusion, supply chain intelligence — supported by advanced technologies and analytics — is a powerful tool for modern procurement and supply chain management. By predicting risks and forecasting demand more accurately, businesses can improve efficiency, reduce costs, and build stronger, more resilient supply chains.

Chapter 7: Predictive Maintenance Of Infrastructure

Let's start with predictive maintenance. This approach focuses on monitoring the condition of equipment to detect early signs of wear and potential failure before they happen. By addressing issues proactively, organizations can reduce unplanned downtime and lower repair costs through timely maintenance. Another critical benefit is improved safety — by preventing unexpected equipment failures, we minimize the risk of hazardous situations for workers and the environment.

Now, what enables predictive maintenance? First, IoT sensors play a crucial role. They collect real-time data from equipment, allowing continuous monitoring and early anomaly detection. Next, machine learning algorithms analyze this data, helping us identify patterns and predict when maintenance will be needed with greater accuracy.

Finally, cloud computing makes it all scalable — it provides the storage and processing power needed to deliver actionable insights in real time.

Predictive maintenance delivers clear benefits like increased operational efficiency and significant cost savings by preventing sudden equipment failures. However, there are challenges. Poor data quality can reduce the effectiveness of these systems, making reliable data collection critical. Integration with existing systems often presents technical hurdles that need careful planning. And, importantly, skilled personnel are required to interpret the data and manage predictive maintenance systems effectively — this often means investing in specialized training.

Infrastructure monitoring relies on various types of sensors. Vibration sensors detect oscillations and movements, which helps in spotting faults or weaknesses early on. Temperature sensors monitor heat fluctuations, safeguarding against material degradation. Strain gauges measure structural deformation, which is key for assessing stress and load-bearing capacity. And acoustic sensors pick up sound waves that can reveal hidden cracks or faults, often before they're visible.

AI processes sensor data using several key methods. Pattern recognition, especially through neural networks, helps identify complex patterns in the data, supporting more accurate decisions. Anomaly detection often uses support vector machines to spot abnormal sensor readings, helping catch problems before they escalate. And failure forecasting, through clustering algorithms, groups similar data patterns to predict potential failures, allowing for proactive maintenance actions.

In real-world applications, AI models excel in two areas: anomaly detection and failure prediction. Anomaly detection helps identify early signs of wear or damage by recognizing unusual behavior patterns in machinery. Failure prediction, on the other hand, uses this early information to enable timely interventions, preventing costly breakdowns and extending the life of assets.

Let's now look at digital twins. These are virtual representations of physical assets, continuously updated with real-time sensor data. They combine this data with engineering models that simulate the behavior and properties of the assets. By applying advanced analytics to digital twins, organizations gain powerful decision support throughout the asset's lifecycle — improving everything from operation to maintenance and replacement planning.

Digital twins integrate seamlessly with predictive maintenance. They enable scenario simulation, allowing organizations to test different conditions and responses virtually. When combined with predictive maintenance, digital twins improve the accuracy of failure predictions. They also help optimize maintenance schedules based on real-time data and simulations, which ultimately enhances asset reliability and performance.

By using simulation and analytics together, digital twins transform asset lifecycle management. They help organizations make smarter decisions by providing a continuous flow of insights from real-time operations, predictive analysis, and virtual testing — all contributing to better planning, cost efficiency, and longer asset life.

To conclude, advanced technologies like predictive maintenance, AI-driven sensor data analysis, and digital twins are reshaping infrastructure management. They offer powerful ways to enhance operational efficiency, reduce costs, and improve safety. By embracing these innovations, organizations can make smarter decisions, optimize maintenance strategies, and ensure the longevity of critical infrastructure assets.

Chapter 8: Defect Detection And Quality Assurance

Defect detection is vital in both manufacturing and construction. By identifying defects early, we prevent compromised structural integrity and ensure high product quality. Common defects we look for include cracks, deformations, and other irregularities that could impact safety and durability. A systematic inspection process helps ensure that every stage of production or construction is thoroughly evaluated, reducing the chance of defects slipping through.

Quality assurance, or QA, is essential for maintaining safety and reliability. QA protocols help ensure that products and structures meet established safety standards, which lowers the risk of failure and protects users. They also maintain product reliability, which in turn strengthens customer trust in the brand. By catching issues early, QA also helps reduce costs related to recalls, repairs, and warranty claims.

Let's compare traditional and advanced methods of defect detection. Traditionally, defect detection has relied on manual inspections and basic imaging techniques. While useful, these methods are slow and prone to human error. In contrast, advanced methods—such as AI and automated sensors—enhance detection accuracy, increase speed, and ensure greater consistency in quality control processes.

Artificial intelligence brings significant improvements to defect detection. Unlike traditional methods, AI can increase both accuracy and speed. Machine learning allows systems to learn from data, automatically identifying defects and adapting to new patterns over time. Computer vision plays a key role here, enabling automated image inspections that can precisely detect anomalies and defects.

There are various techniques for using AI to identify cracks and deformations. These include using image recognition algorithms, machine learning models trained on defect datasets, and advanced computer vision systems. These AI techniques allow for real-time, automated detection that surpasses traditional visual inspections, making the process more reliable and scalable.

Let's look at some practical applications. For bridge inspections, AI systems automate the detection of structural defects, which improves safety and efficiency. In pipeline monitoring, AI tools identify potential defects early, helping to prevent failures and reduce downtime. On manufacturing lines, AI-driven automated inspections improve defect identification, enhance operational efficiency, and contribute to better overall product quality.

Automation plays a crucial role in regulatory compliance. Automated systems enforce standards consistently, which improves reliability across compliance processes. By reducing human error, these systems enhance the accuracy and trustworthiness of compliance checks. Additionally, automation improves transparency by providing clear documentation and tracking, which supports regulatory audits and reporting.

Integrating automated systems with quality assurance protocols creates a continuous monitoring environment. Automated compliance checks ensure that standards are consistently met within QA workflows. Real-time monitoring and reporting give immediate insights, enabling quick issue identification. This integration also supports faster corrective actions and helps drive continuous improvement through data-driven insights.

Automated compliance checks offer clear benefits and present some challenges. On the plus side, they improve efficiency and accuracy by reducing human errors. They also enhance traceability through automated record-keeping, simplifying audits and compliance verification. However, these systems can be complex and costly to implement and maintain. Moreover, they require regular updates and adaptation to keep pace with changing regulations and industry standards.

To conclude, AI-driven defect detection and automated compliance checks are transforming quality assurance. They enhance safety, improve reliability, and increase efficiency across industries. While there are challenges—like system complexity and the need for ongoing updates—the benefits of accuracy, transparency, and continuous improvement make these advanced methods invaluable in today's quality assurance landscape.

Chapter 9: AI For Carbon Emission Reduction

AI plays a key role in identifying the major sources of carbon emissions. With its ability to process vast datasets, AI can highlight key emission sources across industries, making it easier to map out where emissions are coming from—especially within complex supply chains. This information allows businesses to target their efforts more effectively, focusing mitigation strategies where they will have the greatest impact.

Data-driven approaches are essential for accurate carbon footprint analysis. Machine learning algorithms help process complex environmental datasets, enabling precise measurement of carbon emissions. Moreover, AI can predict future emission trends by analyzing both historical and real-time data, giving industries valuable insights for proactive planning and reduction strategies.

Now let's look at some practical examples of how AI is being used to reduce emissions. These include optimizing material choices, predicting environmental impacts during the design phase, and implementing simulation tools that enhance energy efficiency in both building and product design.

AI is transforming how materials are selected and assessed for sustainability. By analyzing material properties through machine learning, AI can identify the best options for specific applications. Additionally, AI uses lifecycle data to assess environmental impacts and durability, which helps in recommending greener, more sustainable materials across industries.

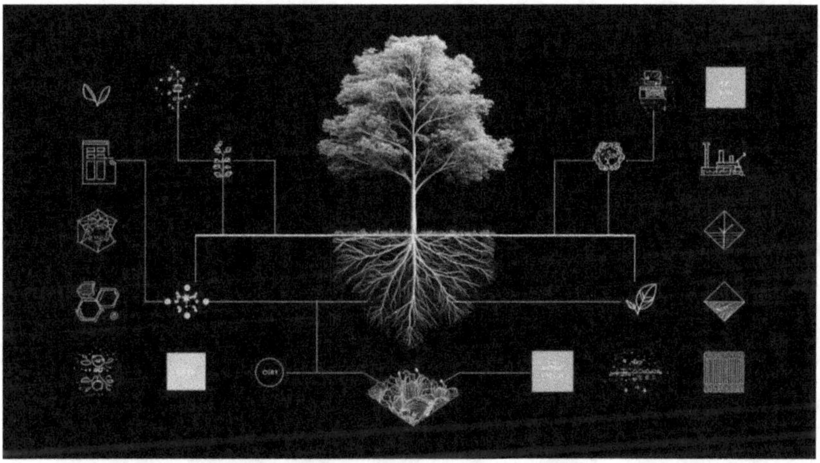

AI also plays a crucial role in predicting the environmental impact of material choices. It can simulate and estimate the carbon footprint of various materials, giving designers and manufacturers the ability to make sustainable decisions right from the early stages of a project. This early analysis helps minimize sustainability risks further down the line.

There are already strong examples of AI helping industries choose more sustainable materials. Whether in construction or the automotive sector, AI has guided the selection of materials that optimize both performance and environmental impact. This not only reduces emissions but also leads to gains in resource efficiency within manufacturing processes.

AI is equally important in the field of energy-efficient design. For architects and product designers, AI enables more efficient designs by optimizing space and enhancing functionality. It also helps minimize energy consumption by promoting better insulation, the smart use of natural resources, and more sustainable design choices overall.

AI-powered simulation tools are incredibly valuable for optimizing energy use. These tools allow designers to model complex energy flows and system efficiencies with great accuracy. By using iterative optimization, they enable multiple rounds of testing and refinement before any actual deployment, ensuring the best possible energy performance.

We've seen many successful implementations of AI-driven strategies that have delivered measurable outcomes. These range from significant carbon emission reductions to improved resource efficiency in various industries, proving that AI can be a powerful tool in driving sustainable innovation.

To conclude, AI offers tremendous potential for reducing carbon emissions through smarter material selection, accurate environmental impact predictions, and optimized design processes. By integrating AI-driven tools and strategies, industries can make meaningful progress towards sustainability goals while also improving efficiency and reducing costs.

Chapter 10: Waste Minimisation And Circular Construction

Let's begin with the principles of waste minimisation in construction. First, efficient design is crucial — when we plan projects with material efficiency in mind, we can prevent a significant amount of waste before construction even begins. Next, material procurement — making sure we only order the materials we actually need helps avoid excess and the associated waste. And finally, recycling practices — by recycling leftover materials on-site or off-site, we not only reduce landfill use but also promote more sustainable construction methods overall.

Now, let's introduce the key concepts of circular construction. Reusing building materials is at the heart of circular construction. Instead of discarding materials, we prioritize finding ways to use them again, which conserves natural resources and minimizes waste. Recycling processes also play a big role — recycling helps us create a closed-loop system, keeping materials in use longer and reducing the need for new resource extraction. And importantly, circular construction considers the entire building lifecycle — from design through to demolition — with sustainability in mind at every stage.

So, what are the benefits of circular construction for sustainability? First, resource efficiency — by reusing and recycling, we make better use of the materials we have. It also helps reduce carbon emissions, both by minimizing waste and by using sustainable materials wherever possible. Cost savings are another big advantage — with efficient material use and less waste to manage, project costs can be significantly reduced. And finally, circular construction drives innovation, encouraging the development of new sustainable techniques and creative building designs.

Let's look at how advanced sorting technologies are improving material reuse on construction sites. AI-driven sorting systems can automatically analyze and categorize materials, improving the precision of sorting. Automation also enhances the separation process, making it faster and reducing the risk of human error. And with sensor-based identification, materials can be accurately sorted by type, which helps divert more waste away from landfills and into reuse or recycling streams.

There are several effective strategies for material categorisation. Clear labeling of materials ensures they can be quickly and correctly identified for reuse or recycling. Establishing on-site sorting protocols gives teams a consistent process to follow, which improves overall sorting efficiency. And, training staff on categorisation techniques is key — this reduces mistakes and helps prevent contamination of recyclable materials, maximizing their value.

We've also seen some great case studies of successful material reuse. One clear benefit is cost reduction — reusing materials means buying less and saving on disposal costs. It also helps reduce environmental impact by conserving resources and cutting down on landfill waste. And importantly, successful projects show that reused materials can still meet or exceed quality standards, ensuring they're safe and durable for use in construction.

Predicting surplus and reducing overordering starts with using data-driven forecasting methods. By analyzing historical data, we can better predict material needs and avoid ordering too much. Machine learning further enhances this by finding patterns in complex data, improving forecasting accuracy. And with predictive analytics, we can optimize our inventory and reduce waste, which ultimately saves costs.

Digital tools play a vital role in inventory management. By using software that tracks inventory in real time, construction teams can plan purchases more accurately, monitor stock levels, and avoid surplus materials sitting unused on-site. This helps ensure that resources are managed efficiently throughout the project.

Best practices for minimizing surplus start with careful planning. Accurate project scoping, clear communication among teams, and regular inventory reviews all contribute to better material control. By adopting these planning practices, construction teams can avoid unnecessary surplus and reduce the risk of waste.

To wrap up, we've explored how waste minimisation and circular construction work together to support sustainable building. From efficient design and smart sorting to predictive analytics and careful inventory management, each of these strategies plays a part in reducing waste and promoting a circular approach. By implementing these practices, we can make construction more sustainable, cost-effective, and innovative.

Chapter 11: Workforce Optimisation

Let's start by understanding workforce optimisation and why it's important. First, workforce alignment—this means ensuring your employees' skills and capacities match your business needs, which is vital for boosting productivity and running operations smoothly. Next, market adaptability—by optimising your workforce, you can quickly adjust to shifting market demands and business environments. And finally, employee engagement—effective optimisation can improve job satisfaction by ensuring people are in roles that fit their skills and strengths.

Workforce optimisation relies on several key components and best practices. Core elements include talent acquisition, scheduling, performance management, and ongoing employee development. Using data analytics helps leaders make better decisions about staffing and development needs. And fostering collaboration and a culture of continuous improvement ensures that optimisation isn't a one-off project, but an ongoing business practice.

So, what are the real benefits of getting this right? Firstly, operational efficiency—better processes and higher productivity. Then, cost reduction—optimising your workforce can significantly lower your operational expenses. You'll also see higher employee satisfaction when people are in roles they enjoy and excel in. And lastly, greater organisational agility, giving your business the ability to pivot quickly when the market shifts.

Now let's look at predicting labour shortages. There are several root causes to understand: Demographic shifts—things like aging populations or shrinking workforces can reduce labour supply. Economic fluctuations—the ups and downs of the economy affect both job availability and labour demand. Skill mismatches—when workers don't have the skills employers need, jobs remain unfilled. And evolving industry demands—as industries change, new skills are needed, often faster than the workforce can adapt.

Data-driven forecasting is essential to predict labour shortages effectively. Looking at historical data helps you identify trends and patterns in labour availability. Understanding labour market trends keeps you informed on current factors shaping workforce supply and demand. And applying predictive analytics allows you to estimate future shortages and plan strategically to avoid disruption.

Proactive workforce planning is key to staying ahead of labour shortages. Scenario analysis lets you plan for different market conditions by anticipating future needs. Succession planning ensures critical roles are covered by developing internal talent pipelines. And adopting flexible workforce strategies means you can adjust staffing levels and skill sets quickly to meet evolving demands.

Now let's move on to skill gap analysis and how AI can support upskilling efforts. First, you need to evaluate current skills to understand your organisation's capabilities. Next, identify skill gaps by comparing those capabilities with what you'll need in the future. Finally, create development initiatives—targeted training programs that build the skills you'll need down the line.

AI plays a powerful role in skill gap analysis. It can quickly process large amounts of data to analyse skill deficiencies at both individual and organisational levels. AI also helps predict emerging skills by tracking industry trends and job market shifts. And it supports personalised learning pathways, ensuring each employee gets training tailored to their specific needs.

When it comes to implementing AI-powered upskilling programs, start with AI-driven workforce analysis to identify skill gaps. Then design tailored learning programs based on AI insights, ensuring relevance and impact. These programs not only meet business needs but also enhance employee engagement and promote continuous growth, aligning development with both individual and organisational goals.

To wrap up—workforce optimisation is about aligning people, processes, and technology for better business outcomes. By predicting labour shortages, leveraging data and AI for skill gap analysis, and implementing targeted upskilling, organisations can improve efficiency, cut costs, boost employee satisfaction, and remain agile in a competitive market. Thank you for watching, and I hope this session has given you valuable insights into using AI for smarter workforce planning.

Chapter 12: AI-Enhanced Risk Management

Traditional risk management faces several core challenges. First, delayed data processing often means that risks are identified too late to mitigate effectively. Second, assessments are frequently subjective—relying on human judgment—which introduces inconsistency and bias. Finally, traditional methods struggle with complex data, which leads to important risk signals being overlooked or underestimated.

Artificial Intelligence introduces powerful capabilities in risk management. Machine learning automates both risk detection and assessment, making processes faster and more accurate. Predictive analytics allow organizations to forecast potential risks based on emerging data patterns. Overall, AI enhances detection by identifying risks earlier and assessing them with greater precision than manual methods.

AI-driven risk assessment delivers several key benefits. With real-time monitoring, risks can be tracked continuously, allowing immediate responses. AI's analytical power improves the accuracy of risk assessments by handling large, complex datasets. This enables proactive mitigation—predicting problems before they occur. Finally, AI helps optimize resource allocation, ensuring that efforts are focused on the most critical risks, reducing both exposure and costs.

AI algorithms evaluate a wide range of project risk factors. They analyze financial data to assess monetary risks and determine if a project is viable. They also review resource availability to foresee constraints or delays. Importantly, they monitor external conditions—like market trends or environmental factors—which helps keep risk predictions accurate and up-to-date.

Real-time risk scoring relies on integrating various live data sources. By continuously feeding fresh data into the AI system, organizations can keep their risk assessments current. This integration allows the AI to dynamically adjust risk scores as new information becomes available, making the scoring system both responsive and relevant.

Let's look at some practical examples. AI-driven risk scoring helps identify project risks early, giving teams the chance to intervene before problems escalate. These proactive interventions can stabilize projects and prevent costly disruptions. Ultimately, using AI in this way leads to reduced delays and lower costs, driving efficiency and improving project outcomes.

Staying compliant with regulations is an ongoing challenge. Regulatory frameworks are constantly evolving, and organizations need to keep pace to avoid penalties. Understanding your compliance obligations is vital to managing risk effectively. Early warning systems support this by flagging compliance risks before they turn into serious violations.

AI-powered early warning systems work by continuously analyzing compliance data. They use anomaly detection to spot irregularities, allowing organizations to catch potential issues early. These systems also generate timely alerts, giving decision-makers the opportunity to act before risks become critical, supporting stronger governance and proactive risk management.

For effective implementation, AI must integrate smoothly with existing business processes. This ensures minimal disruption and maximum benefit. Training is also critical—your team needs the skills to work confidently with AI systems. Finally, AI applications span sectors like finance, healthcare, and manufacturing, where they not only improve operations but also significantly reduce compliance risks.

To conclude, AI-enhanced risk management transforms how organizations approach risk and compliance. By leveraging AI's capabilities—like real-time monitoring, predictive analytics, and early warning systems—businesses can move from reactive to proactive risk strategies. This leads to better decision-making, fewer disruptions, and stronger compliance in a rapidly changing world.

Chapter 13: Human-AI Collaboration In Engineering

Let's start by looking at how humans and AI complement each other in engineering partnerships. Humans contribute creativity, critical thinking, and domain expertise—qualities essential for innovation. On the other hand, AI excels at processing large datasets, finding patterns, and offering insights that would be difficult for humans to detect alone. When combined, these strengths lead to smarter, more efficient problem-solving in engineering projects.

However, collaboration isn't without its challenges. On the positive side, human-AI workflows can enhance both productivity and accuracy. But building trust in AI systems and understanding how they make decisions remain significant hurdles. And integrating AI into existing engineering workflows often requires careful planning and technical adjustments, which can be complex.

Now, let's explore the types of AI-augmented decision-making tools commonly used in engineering. Machine learning models help analyze data patterns to improve decision accuracy. Expert systems replicate human expertise for tackling complex problems. Decision support systems provide actionable insights to assist engineers in making informed choices. And predictive analytics platforms forecast outcomes to help optimize project results.

AI enhances human judgment in several key ways. First, it increases data analysis efficiency, allowing engineers to quickly access and interpret large amounts of information. Second, AI identifies patterns in data that might be missed by humans, leading to more accurate problem-solving. Third, AI can simulate various scenarios, helping engineers anticipate possible outcomes and make better-informed decisions.

Let's look at some examples of these tools in action. AI-assisted CAD systems streamline design work and spark creativity. Structural health monitoring systems use AI to detect potential issues before they become serious, improving safety and reducing maintenance costs. Automated code compliance tools help ensure designs meet regulatory standards, reducing errors and enhancing project reliability.

Moving on to ethics in AI-driven design. First, we must address bias—AI systems need to be carefully managed to prevent biased outcomes that could affect fairness. Second, privacy protection is critical; engineering solutions must respect user data and comply with legal standards. Third, safety must always be a priority in AI design to protect both users and the public. Lastly, we have to consider AI's broader societal impacts to ensure responsible use.

Transparency and accountability are key to building trust in AI systems. Transparency means making AI decision-making processes understandable and open to scrutiny. Accountability mechanisms must be in place to address any mistakes or adverse consequences that arise from AI use. Together, transparency and accountability help foster user confidence in AI-assisted engineering decisions.

Responsible AI engineering also involves following best practices and frameworks. Adopting industry standards ensures AI systems are reliable and consistent. Implementing ethical guidelines promotes fairness and accountability. And conducting interdisciplinary reviews—bringing in diverse perspectives—helps identify both technical and social implications of AI, promoting well-rounded, responsible solutions.

To conclude, human-AI collaboration in engineering holds immense potential for innovation and improved outcomes. But success depends on thoughtful integration, trust, ethical responsibility, and a commitment to transparency and accountability. By working together—humans and AI—we can unlock new levels of creativity, efficiency, and positive impact in engineering.

Chapter 14: Data Governance And AI Standards In Construction

Data governance in construction refers to the policies, processes, and standards that control how data is managed across projects. This means ensuring that project data is accurate, accessible, secure, and used responsibly throughout the lifecycle of a construction project. Good governance helps teams make informed decisions and maintain compliance with regulations.

AI standards play several important roles in construction project management and operations. They ensure AI systems are reliable, work seamlessly with other tools, and comply with industry requirements. For project scheduling, AI can help manage tasks and timelines more effectively. AI also improves safety by detecting hazards and alerting teams in real time. And when it comes to resource allocation, AI standards help ensure materials, equipment, and labor are used efficiently.

Implementing data governance and AI standards in construction comes with challenges. Data sources are often fragmented, coming from various teams and systems that don't always integrate easily. Different groups adopt technology at different speeds, leading to inconsistency. And because construction sites change constantly, it's difficult to apply standardized approaches to data and AI governance in such dynamic environments.

Construction sites collect a wide variety of data. This includes employee information for workforce management and safety, tracking how equipment is used for maintenance and efficiency, surveillance data for security, project progress metrics to ensure timely delivery, and environmental monitoring data to meet safety and compliance standards. Each type of data plays a vital role in site operations.

To protect site data, several best practices should be followed. Access controls help make sure only authorized personnel can access sensitive information. Encrypting data means even if it's intercepted, it remains secure. Continuous monitoring helps detect suspicious activities early. And training employees ensures everyone understands how to protect data and avoid accidental breaches.

Construction data systems face several risks. Cyberattacks can exploit system vulnerabilities, especially if software isn't updated. Insider threats—whether accidental or deliberate—can also compromise data. Learning from real-world data breaches helps companies identify weaknesses and strengthen their security measures, preventing future incidents.

Several regulations impact AI use in construction. Data protection laws require personal and sensitive information to be handled securely. Safety standards guide the use of AI to ensure it helps protect workers and prevent hazards. And industry guidelines set expectations for ethical, effective AI deployment on construction projects.

Key standards like those from ISO and IEEE provide frameworks to guide AI system development, focusing on reliability, transparency, and accountability. In construction, firms must comply with these standards to ensure AI systems are safe and effective. The most important compliance areas are ensuring AI systems are reliable, operate transparently, and can be held accountable.

Emerging regulatory trends emphasize the responsible use of AI in construction, promoting safety and reliability. Ethical considerations like fairness and accountability are increasingly central to new regulations. Additionally, transparency in how AI systems operate and cooperation between global regulators are essential to create effective, consistent frameworks worldwide.

To wrap up, integrating data governance, privacy and security measures, and compliance with regulations is crucial for the successful use of AI in construction. By addressing these areas together, companies can harness AI's potential while safeguarding their projects, workforce, and data.

www.ingramcontent.com/pod-product-compliance
Lightning Source LLC
LaVergne TN
LVHW022316200926
848481LV00004B/1364

* 9 7 9 8 8 9 8 6 0 3 8 0 9 *